CONSERVATION CHAMPIONS!

COMPOSTING

SADIE SILVA

Enslow
PUBLISHING

Please visit our website, www.enslow.com.
For a free color catalog of all our high-quality books, call toll free 1-800-398-2504 or fax 1-877-980-4454.

Library of Congress Cataloging-in-Publication Data
Names: Silva, Sadie author
Title: Composting / Sadie Silva.
Description: Buffalo, NY : Enslow Publishing, [2027] | Series: Conservation Champions! | Includes index. | Audience: Grades 2-3 | Summary: "This book explains how foods can be composted and why compost is good for the environment"– Provided by publisher.
Identifiers: LCCN 2025052936 (print) | LCCN 2025052937 (ebook) | ISBN 9781978547681 library binding | ISBN 9781978547674 paperback | ISBN 9781978547698 ebook
Subjects: LCSH: Compost–Juvenile literature
Classification: LCC TD796.5 .S55 2027 (print) | LCC TD796.5 (ebook) | DDC 631.8/75–dc23/eng/20260202
LC record available at https://lccn.loc.gov/2025052936
LC ebook record available at https://lccn.loc.gov/2025052937

Published in 2027 by
Enslow Publishing
2544 Clinton Street
Buffalo, NY 14224

Designer: Tanya Dellaccio Keeney
Editor: Therese Shea

Photo credits: Cover (person) Halfpoint/Shutterstock.com; cover (garbage), p. 9 Pixel-Shot/Shutterstock.com; p. 5 tanitost/Shutterstock.com; p. 7 K-FK/Shutterstock.com; p. 11 (right) tawanroong/Shutterstock.com; p. 11 (left) Thaisucculents/Shutterstock.com; p. 13 (right) grandbrothers/Shutterstock.com; p. 13 (left) nieriss/Shutterstock.com; p. 15 Mariana Serdynska/Shutterstock.com; p. 17 Jub Jub Photography/Shutterstock.com; p. 19 (left) Here Now/Shutterstock.com; p. 19 (right) HM Scott/Shutterstock.com; p. 21 wavebreakmedia/Shutterstock.com.

Printed in the United States of America

Some of the images in this book illustrate individuals who are models. The depictions do not imply actual situations or events.

CPSIA compliance information: Batch #CSENS27: For further information contact Enslow Publishing, at 1-800-398-2504.

CONTENTS

BOLDFACE WORDS APPEAR IN THE GLOSSARY.

DON'T THROW IT OUT!

What do you do with apple cores and banana peels? Do you put them in the trash? They don't have to be waste. They can be used to make compost. Compost creates beautiful gardens! Composting is a way to help Earth. Read on to learn how.

COMPOST IS A MIX OF ROTTING PLANTS AND OTHER THINGS. IT'S ADDED TO SOIL, OR DIRT. MAKING THIS MIX IS CALLED COMPOSTING.

COMPOSTING IS RECYCLING

Recycling is turning old or used things into something new. People who compost turn some kinds of waste into a matter that **improves** soil. Composting keeps this waste from ending up in a landfill. A landfill is where trash is taken to decompose, or break down, over time.

MANY THINGS TAKE A LONG TIME TO DECOMPOSE IN A LANDFILL.

LESS IN LANDFILLS

We need to put trash somewhere! But landfills have problems. They give off harmful gases. They **pollute** water and soil if harmful **liquids** leak out. They draw animals who get sick when they eat harmful things in landfills. It's good to keep as much out of landfills as we can.

COMPOSTING IS A GREAT WAY TO CUT DOWN ON LANDFILL WASTE.

WHAT'S IN IT?

Not all waste can be turned into compost. You need the right mix. Things that were once plants are good for compost, such as vegetables, fruits, and cut grass. They're often a bit wet. You also need dry materials, or matter, such as dead leaves, wood bits, and paper.

EGGSHELLS ARE ANOTHER GOOD THING FOR COMPOST. SO ARE COFFEE GROUNDS AND PET HAIR!

WHERE TO PUT IT

The compost mix should go in a bin or box with a top on it, so pests can't get at it. It should be kept **moist**. If it's too wet, more dry materials should be added. Each time materials are added to it, the compost should be mixed up. This lets air in.

COOKED FOOD, BONES, MEAT, FISH, OILS, AND MILK PRODUCTS AREN'T GOOD FOR COMPOST. THEY SMELL BAD AND DRAW PESTS SUCH AS RATS!

A BIG CHANGE

After a few months, the parts of the compost mix decompose. The compost starts to look dark brown. It won't smell like rotting food. It will smell like soil. **Organisms** such as worms and **bacteria** help compost change form.

WHEN COMPOST TURNS INTO THIS DARK FORM, IT'S CALLED HUMUS.

WHY ADD COMPOST?

Compost can be added to garden soil. It gives garden plants **nutrients** they need. It helps keep the soil moist in hot weather. It helps the soil stay a good **temperature**. Finally, it helps stop weeds from growing!

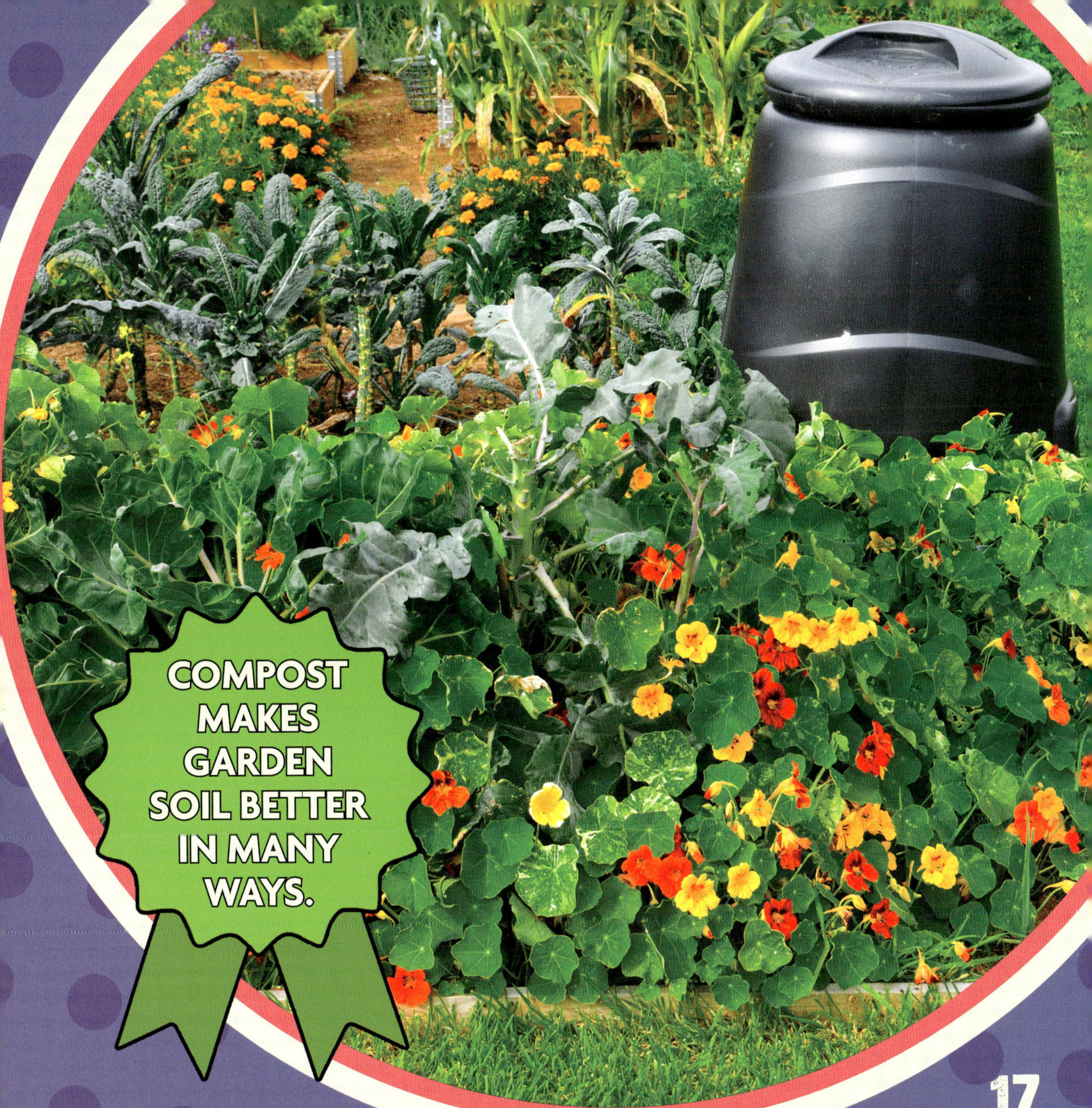

COMPOST MAKES GARDEN SOIL BETTER IN MANY WAYS.

CARING ABOUT CONSERVATION

Conservation means caring for Earth's **resources**. Cutting down on waste in landfills is a form of conservation. So is composting waste and using it to grow your own food and flowers. There are many ways to be a conservation champion!

SOME CITIES COLLECT WASTE TO COMPOST.

ACTIVITY

- For one day, keep track of what is thrown in the trash at home. Write it down as a list.
- Put a mark next to everything that could be composted.
- Show an adult in your home your list.
- Ask if you can start composting at home. If not, find out if you can join a town or city composting program.

COMPOST

GLOSSARY

bacteria: Tiny living things that can only be seen with a microscope.

improve: To make something better.

liquid: Something that flows and takes the shape of the container holding it.

moist: Having some water.

nutrient: Something a living thing needs to grow and stay alive.

organism: A living thing.

pollute: To become unsafe for people or animals.

resource: A usable supply of something.

temperature: How hot or cold something is.

FOR MORE INFORMATION

BOOKS

Chu, Katherine. *Composting Fun*. Minneapolis, MN: Lerner Publications, 2025.

Hansen, Amy. *Curious about Composting*. Mankato, MN: Amicus Learning, 2025.

WEBSITES

Create a Compost
kids.nationalgeographic.com/books/article/create-compost
Follow these steps to create your own compost pile.

Compost Facts for Kids
kids.kiddle.co/Compost
Check out more facts about the benefits of composting.

Publisher's note to educators and parents: Our editors have carefully reviewed these websites to ensure that they are suitable for students. Many websites change frequently, however, and we cannot guarantee that a site's future contents will continue to meet our high standards of quality and educational value. Be advised that students should be closely supervised whenever they access the internet.

INDEX